1 Counting
1

To parents
Please remove the first sheet of stickers and give it to your child. After pasting a sticker, it may be good for you to read the numbers at the bottom of the page aloud to your child while pointing with your finger. The numbers in the yellow squares are numbers your child has already learned in this book.

Paste the sticker as you like.
Then, count the dog as " 1 " while pointing to the image.

| 1 | 2 | 3 | 4 | 5 | 6 | 7 | 8 | 9 | 10 |

Counting
I

To parents

After pasting a sticker, it is good to say "one" with your child while pointing to the number "1" at the bottom of the page with your finger. It is also good to count the cat picture and black dots at the bottom of the page.

Count the number of red dots (🔴) while pointing to each one. Then, paste the sticker on the dot.

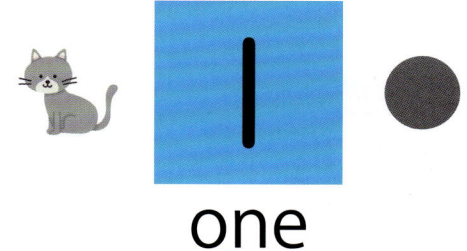

one

3 **Counting** 1, 2

To parents

When your child is finished, talk with him or her about the picture. You can say something like, "This is a fish. Fish swim in the sea."

Paste the stickers as you like.
Then, count the fish while pointing to each one.

1 2 3 4 5 6 7 8 9 10

Counting
1, 2

To parents

When pasting the stickers on each red circle, it is okay if the stickers do not match up perfectly. Let your child enjoy the fun of pasting the sticker on his or her own.

Count the number of red dots(🔴) while pointing to each one. Then, paste a sticker onto each dot.

two

5 Counting 1-3

To parents

If your child peels off the sticker in a hurry, it could rip. Please assist him or her by saying something cautious like, "Please peel the sticker off slowly." It is important you let your child practice peeling off and placing the stickers to help his or her fine motor development.

Paste the stickers as you like.
Then, count the elephants while pointing to each one.

| 1 | 2 | 3 | 4 | 5 | 6 | 7 | 8 | 9 | 10 |

6 Counting 1-3

To parents

Peeling off and pasting tiny stickers is not an easy skill for young children to learn. Your child must gain the ability to control the fine movements of his or her fingers. When your child is finished with the activity, please offer lot of praise.

Count the number of red dots(⬤) while pointing to each one. Then, paste a sticker onto each dot.

 **3**

three

Counting
1-4

To parents

It is okay if your child pastes stickers for apples outside of the tree, you do not have to have your child fix them. Let him or her have fun while you enjoy watching your child's progress.

Paste the stickers as you like.
Then, count the apples while pointing to each one.

| 1 | 2 | 3 | 4 | 5 | 6 | 7 | 8 | 9 | 10 |

Counting 1-4

To parents

If your child seems to have difficulty counting the red dots while pointing to them one by one with his or her finger, you may point to the dots with your finger instead.

Count the number of red dots(🔴) while pointing to each one. Then, paste a sticker onto each dot.

4

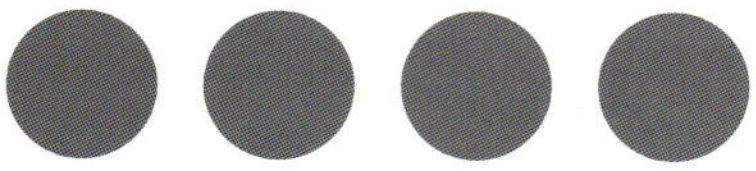

four

Counting
1-5

To parents
It is okay if your child pastes a car sticker on the lines on the road. What is most important is that your child enjoys stickering.

Paste the stickers as you like.
Then, count the cars while pointing to each one.

1	2	3	4	5	6	7	8	9	10

Counting
1-5

To parents

When your child can count the number, offer lots of praise. It would be a good idea to tell your child how to read numbers while pointing at the number with your finger.

Count the number of red dots(●) while pointing to each one. Then, paste a sticker onto each dot.

5

five

11 Counting
1-5

To parents

Please peel off the stickers in turn from 1 to 5 and give each one to your child. It is not necessary to change it even if your child pastes the sticker at a different position than the hint shown on this page. Guide your child to paste stickers and count them with fun.

Count the balloons while pasting each one in the picture.

Counting
1-5

To parents

Please peel off the sticker in turn from 1 to 5 and give it to your child. If your child does not understand where the stickers should be pasted, you should tell your child where to paste them.

Read the numbers while pointing to each one.

1	2	3	4	5

Read the numbers while pasting each sticker.

1	2	3	4	5

Numbers
1-5

To parents

After finishing the activity on this page, it is good to count the number of concrete objects and red dots by pointing with your finger. If your child can count the numbers well, offer lots of praise.

Paste the stickers and say the numbers aloud.

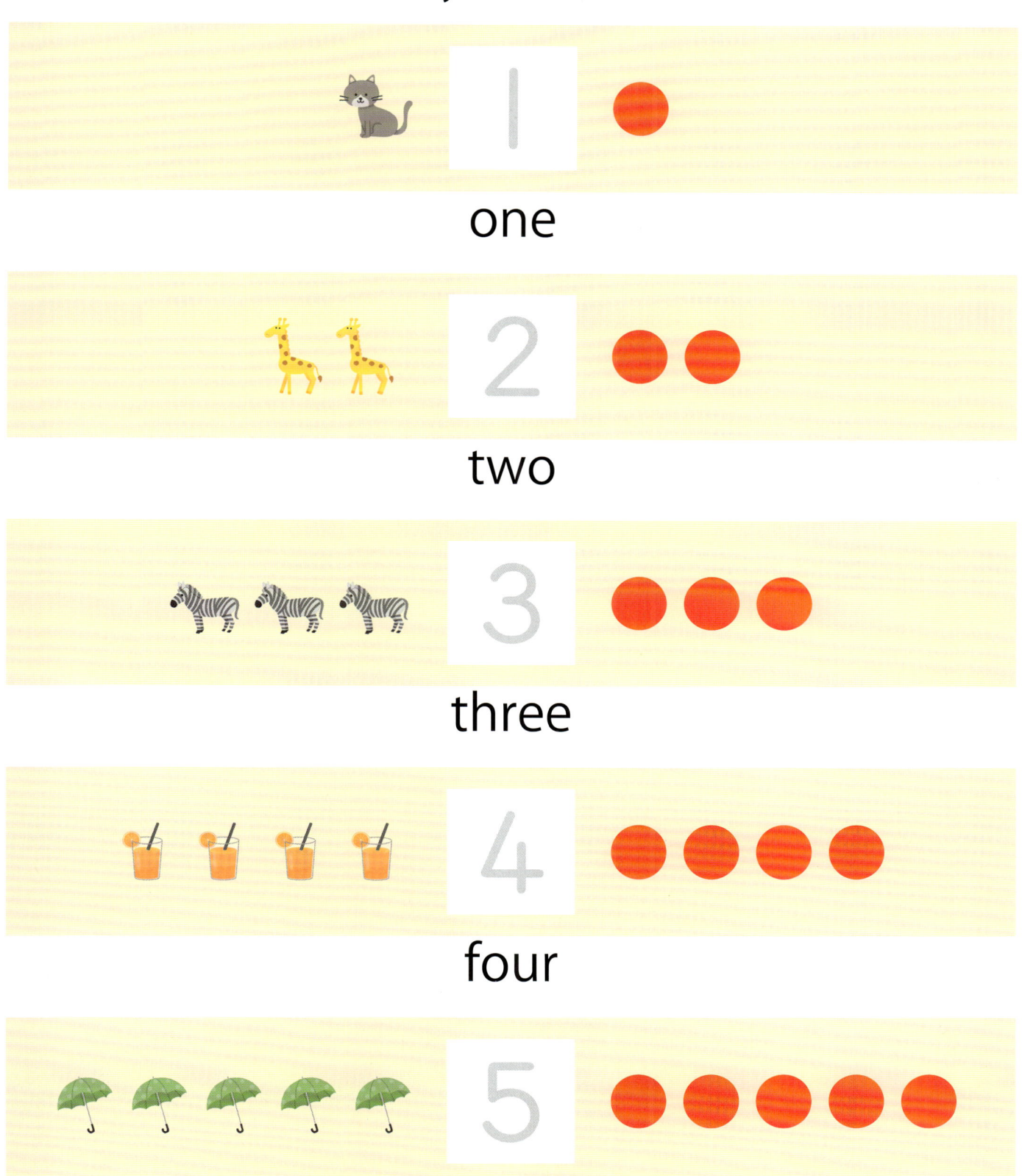

one

two

three

four

five

Numbers
1-5

To parents

This activity is meant to teach your child to correctly distinguishing numbers and colors, while completing the picture with stickers. This may be a difficult activity for your child. It would be good to help him or her by saying something like, "Let's paste a pink sticker on the triangle that says 1".

Paste the stickers on the correct numbers to create a picture.

1 = **Pink** / 3 = **Red** / 4 = **Yellow**

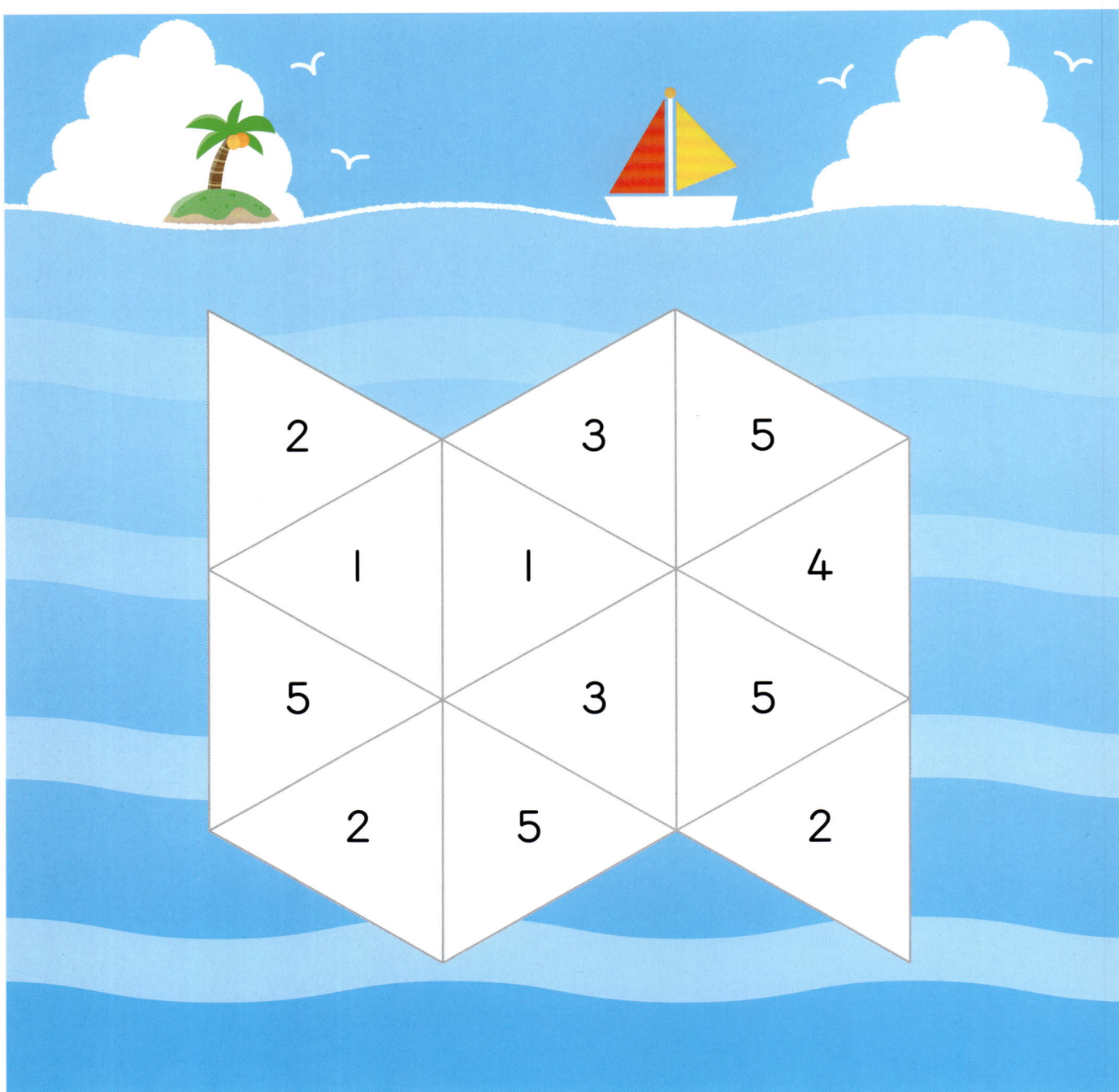

Counting
1-6

To parents

After pasting a sticker, it may be good for you to read the numbers at the bottom of the page aloud to your child while pointing with your finger.

Paste the stickers as you like.
Then, count the doughnuts while pointing to each one.

1	2	3	4	5	6	7	8	9	10

Counting
1-6

To parents

After pasting a sticker, it may be good to say "six" with your child while pointing to the number "6" at the bottom of the page with your finger. It is also good to count the bus pictures and black dots at the bottom of the page for more practice.

Count the number of red dots(●) while pointing to each one. Then, paste a sticker onto each dot.

six

Stickers - 1

To be used in 1

To be used in 2

To be used in 3

To be used in 4

To be used in 5

To be used in 6

To be used in 7

To be used in 8

Stickers-2

To be used in 9

To be used in 10

To be used in 12

To be used in 11

To be used in 13

Stickers -3

To be used in 15

To be used in 14

4

1

To be used in 17

To be used in 16

To be used in 18

To be used in 19

To be used in 20

Stickers -4

To be used in 22

To be used in 21

CRAYON

CRAYON

CRAYON

CRAYON

CRAYON

CRAYON

CRAYON

CRAYON

To be used in 23

To be used in 24

To be
used in 25

6 7 8 9 10

To be used in 26

6 7 8 9 10

To be used in 27

6 7 8 9 10

To be used in 28

7 7 7 7 10 10 9

To be used in 29

1 2 3 4 5

6 7 8 9 10

Counting
1-7

To parents
It may be good to count the stickers as "one, two" while pasting them. When your child is finished, talk with him or her about the picture. You can say something like, "These are tulips. What color tulips do you see?"

Paste the stickers as you like.
Then, count the tulips while pointing to each one.

1	2	3	4	5	6	7	8	9	10

Counting
1-7

To parents
If your child seems to have difficulty counting the red dots while pointing to them one by one with his or her finger, you may point to the dots with your finger instead as your child counts aloud.

Count the number of red dots(⬤) while pointing to each one. Then, paste a sticker onto each dot.

seven 7

Counting 1-8

To parents

It is okay if your child pastes stickers for crayons outside of the case, you do not have to have your child fix them. Let him or her have fun while you enjoy watching your child's progress.

Paste the stickers as you like.
Then, count the crayons while pointing to each one.

| 1 | 2 | 3 | 4 | 5 | 6 | 7 | 8 | 9 | 10 |

Counting
1-8

To parents

It may be good to count the butterflies as "one, two" while pasting the stickers. When pasting the stickers on each red circle, it is okay if the stickers do not match up perfectly.

Count the number of red dots() while pointing to each one. Then, paste a sticker onto each dot.

8
eight

Counting
1-9

To parents

Some small stickers are used on this page. Peeling off a small sticker by himself or herself may be a difficult activity for your child. When your child is finished, please offer lots of praise.

Paste the stickers as you like.
Then, count the penguins while pointing to each one.

| 1 | 2 | 3 | 4 | 5 | 6 | 7 | 8 | 9 | 10 |

Counting
1-9

To parents
Even if your child paste stickers that overlap with other stickers, you do not have to have your child fix them. If your child peels off the sticker after it is placed, it could rip, so parents should watch and help him or her be careful.

Count the number of red dots() while pointing to each one. Then, paste a sticker onto each dot.

9

nine

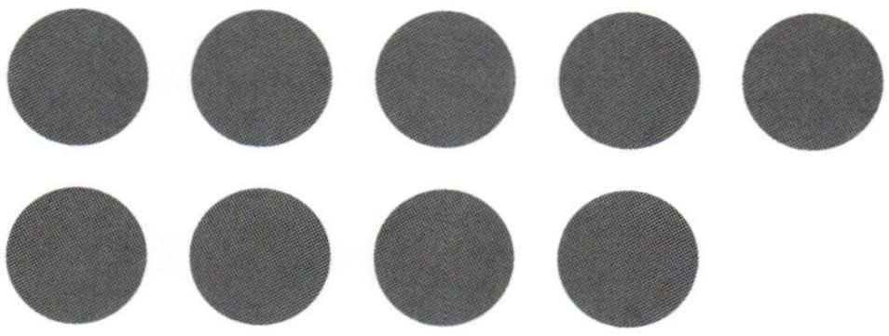

Counting
1-10

To parents

There are various types of airplane stickers on this page. Your child does not need to worry about how to paste the stickers. What is most important is that your child enjoys stickering as he or she counts.

Paste the stickers as you like.
Then, count the airplanes while pointing to each one.

1	2	3	4	5	6	7	8	9	10

24 Counting
1-10

To parents

When your child can count up to the number on the page, offer lots of praise. It would be a good idea to teach your child how to read the numbers while pointing at each number with your finger.

Count the number of red dots( ) while pointing to each one. Then, paste a sticker onto each dot.

10
ten

25 Counting 1-10

To parents

Please peel off the stickers in turn from 6 to 10 and give each one to your child. It is not necessary to change it even if your child pastes the sticker at a different position than the hint shown on this page. Guide your child to paste stickers and count them with fun.

Count the boats while pasting each one in the picture.

Counting
1-10

To parents
Please peel off the sticker in turn from 6 to 10 and give it to your child. If your child does not understand where the stickers should be pasted, you should help your child find where to paste them.

Read the numbers while pointing to each one.

1	2	3	4	5
6	7	8	9	10

Read the numbers while pasting each sticker.

1	2	3	4	5
6	7	8	9	10

Numbers
6-10

To parents

After finishing the activity on this page, it is good to count the number of concrete objects and red dots by pointing with your finger. If your child can count the numbers well, offer lots of praise.

Paste the stickers and say the numbers aloud.

six

seven

eight

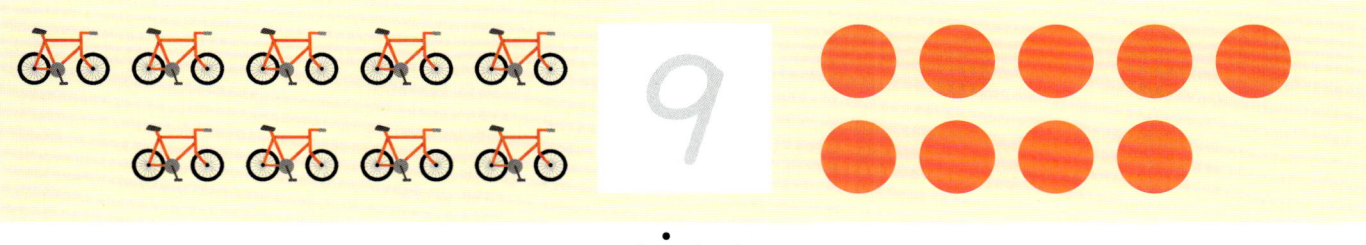

nine

ten

Numbers
1-10

To parents
This activity is meant to teach your child to correctly distinguish numbers and colors, while completing the picture with stickers. This may be a difficult activity for your child. It would be good to help him or her by saying something like, "Let's paste a red sticker on the triangle that says 7".

Paste the stickers on the correct numbers to create a picture.

7 = **Red** / 9 = **Yellow** / 10 = **Blue**

Review

To parents

This is the last exercise in this workbook. Offer your child lots of praise along with the Certificate of Achievement on the end of the workbook. Let's count the various things around us so as to further increase the number sense of your child.

Count the pictures and red dots in each frame.
Then, say each number aloud as you paste the stickers.

page 1

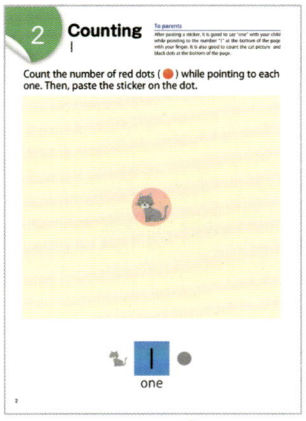

page 2

page 3

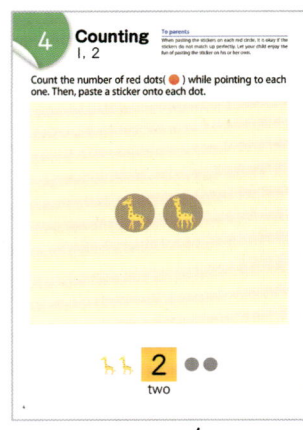

page 4

page 5

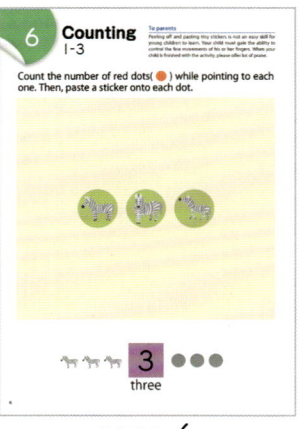

page 6

page 7

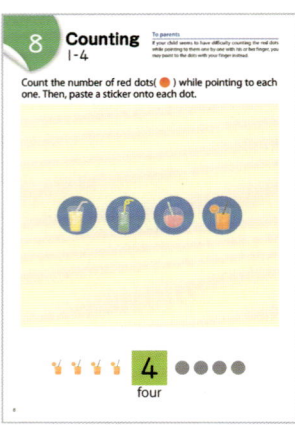

page 8

page 9

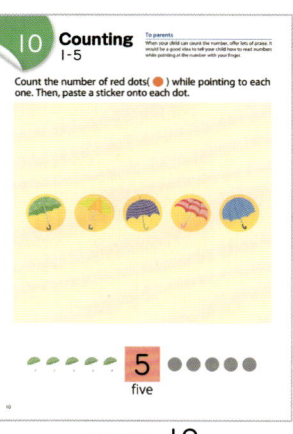

page 10

page 11

page 12

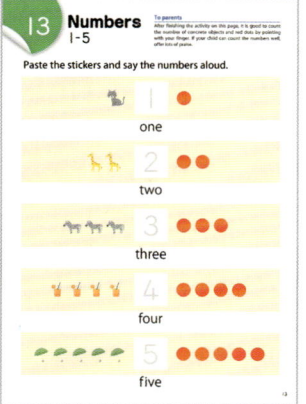

page 13

page 14

page 15

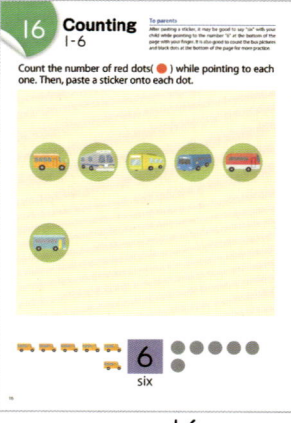

page 16

page 17

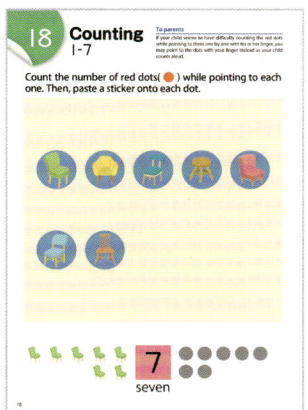

page 18

page 19

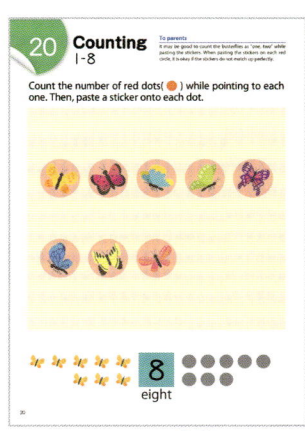

page 20

page 21

page 22

page 23

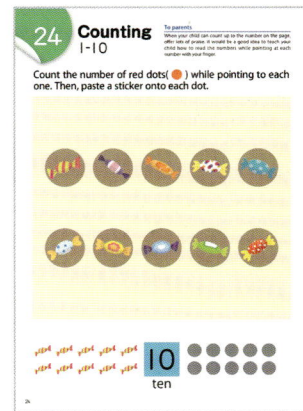

page 24

page 25

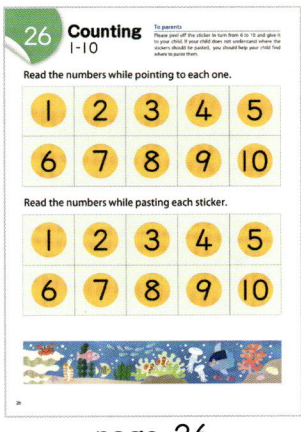

page 26

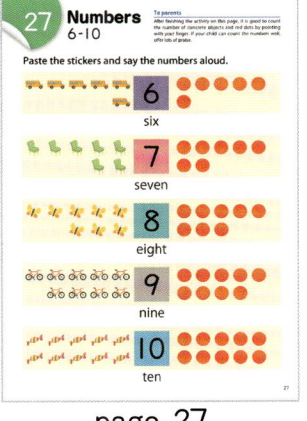

page 27

page 28

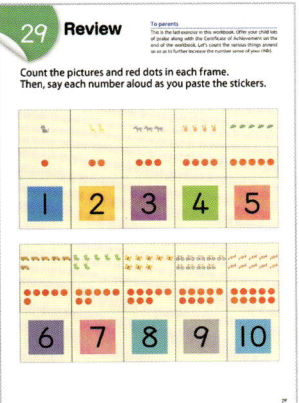

page 29

Certificate of Achievement

Good job!

_____

is hereby congratulated on completing
Counting with Stickers
1-10

Presented on _____ , 20 _____

_____

Parent or Guardian